YOUR KNOWLEDGE HAS VALUE

- We will publish your bachelor's and master's thesis, essays and papers

- Your own eBook and book - sold worldwide in all relevant shops

- Earn money with each sale

Upload your text at www.GRIN.com and publish for free

Bibliographic information published by the German National Library:

The German National Library lists this publication in the National Bibliography; detailed bibliographic data are available on the Internet at http://dnb.dnb.de .

This book is copyright material and must not be copied, reproduced, transferred, distributed, leased, licensed or publicly performed or used in any way except as specifically permitted in writing by the publishers, as allowed under the terms and conditions under which it was purchased or as strictly permitted by applicable copyright law. Any unauthorized distribution or use of this text may be a direct infringement of the author s and publisher s rights and those responsible may be liable in law accordingly.

Imprint:

Copyright © 2017 GRIN Verlag, Open Publishing GmbH
Print and binding: Books on Demand GmbH, Norderstedt Germany
ISBN: 9783668486584

This book at GRIN:

http://www.grin.com/en/e-book/369798/phytochemical-analysis-of-fruit-extracts-of-baccaurea-courtallensis-and

Prem Jose Vazhacharickal, Jiby John Mathew, Sajeshkumar N. K., Esther Elsa George

Phytochemical analysis of fruit extracts of Baccaurea courtallensis and evaluation of cholesterol lowering property

GRIN Publishing

GRIN - Your knowledge has value

Since its foundation in 1998, GRIN has specialized in publishing academic texts by students, college teachers and other academics as e-book and printed book. The website www.grin.com is an ideal platform for presenting term papers, final papers, scientific essays, dissertations and specialist books.

Visit us on the internet:

http://www.grin.com/

http://www.facebook.com/grincom

http://www.twitter.com/grin_com

Phytochemical analysis of fruit extracts of *Baccaurea courtallensis* and evaluation of cholesterol lowering property

Prem Jose Vazhacharickal, Jiby John Mathew, Sajeshkumar N.K and Esther Elsa George

ACKNOWLEDGEMENTS

Firstly we thank **God Almighty** whose blessing were always with us and helped us to complete this project work successfully.

We wish to thank our beloved Manager **Rev. Fr. Dr. George Njarakunnel,** Respected Principal **Dr. Joseph V.J,** Vice Principal **Fr. Joseph Allencheril**, Bursar **Shaji Augustine** and the Management for providing all the necessary facilities in carrying out the study. We express our sincere thanks to **Mr. Binoy A Mulanthra** (lab in charge, Department of Biotechnology) for the support. This research work will not be possible with the co-operation of many farmers.

Lastly, we extend our indebt thanks to patents, friends, and well wishers for their love and support.

Prem Jose Vazhacharickal*, Jiby John Mathew, Sajeshkumar N.K and Esther Elsa George

*Address for correspondence
Assistant Professor
Department of Biotechnology
Mar Augusthinose College
Ramapuram-686576
Kerala, India

Table of contents

Table of contents..iii

Table of figures iv

Table of tables...iv

List of abbreviations iv

Phytochemical analysis fruit extracts of *Baccaurea courtallensis* and evaluation of
cholesterol lowering property 1

Abstract.. 1

1. Introduction 2

    1.1 Objectives 3

2. Review of literature.. 3

3. Hypothesis 6

4. Materials and Methods 6

    4.1 Study area.. 6

    4.2 Collection of plant material.. 6

    4.3 Preparation of *Baccaurea courtallensis* fruit pulp extracts 6

    4.4 Phytochemical screening 6

    4.5 Preparation of cholesterol samples 9

    4.6 Treatment.. 9

    4.7 Estimation of cholesterol 9

    4.9 Statistical analysis... 9

5. Results and discussion... 16

6. Conclusions... 17

Acknowledgements 17

References.. 18

Table of figures

Figure 1. Map of Kerala showing the soil sample collection point. 10

Figure 2. Details of the *Baccaurea courtallensis* plant with fruits. 11

Figure 3. Details of phytochemical analysis of *Baccaurea courtallensis* fruit (water extract S1); A. terpenoids B. anthocyanins C. carbohydrates D. saponins E. phlobatanins F. glycosides G. anthraquinones H. emodins I. steroid J. leucoanthocyanin K. coumarins L. proteins M. flavonoids N. phenols O. alkaloids. . 12

Figure 4. Standard graph for cholesterol estimation by Zak's method. 14

Table of tables

Table 1. Preliminary phytochemical analysis of *Baccaurea courtallensis* fruit extracts. 13

Table 2. Cholesterol estimation at different time intervals after treatment (n=3; values in mg/g sample).. 15

List of abbreviations

%	: Percentage
°C	: Degree celsius
µL	: Microlitre
AOAC	: Association of official analytical chemistry
CVD	: Cardio vascular disease
HDL	: High density lipoprotein
LDL	: Low density lipoprotein
TAG	: Triglycerides

Phytochemical analysis fruit extracts of *Baccaurea courtallensis* and evaluation of cholesterol lowering property

Prem Jose Vazhacharickal[1]*, Jiby John Mathew[1], Sajeshkumar N.K[1] and Esther Elsa George[1]

[1]Department of Biotechnology, Mar Augusthinose College, Ramapuram, Kerala, India-686576

Abstract

The experiment was carried out to extract and analyze the phytochemical constituents of the *Baccaurea courtallensis* fruit and to find out the cholesterol lowering efficacy of the extract. The water extracts of *Baccaurea courtallensis* fruits were subjected to preliminary phytochemical analysis and they showed the presence of alkaloids, flavonoids, terpenoids, saponins, phlobatannins, coumarin, anthocyanin, leucoanthocyanin, phenols and carbohydrates. The extract was evaluated for cholesterol lowering efficiency against different fatty food materials like egg yolk, pork and chicken fat, ghee and cod liver oil by Zak's method. The maximum efficiency was observed on egg yolk and chicken fat followed by pork fat and ghee. In cod liver oil no beneficial change were noticed.

Keywords: Cholesterol; Zak's method; Hypercholesterolemia, Antihyperlipidemic, Emodins, Coumarins.

1. Introduction

Hundreds of fruit-bearing trees are native to Southeast Asia, but many of them are considered as indigenous or underutilized. These species can be categorized as indigenous tropical fruits with potential for commercial development and those possible for commercial development. Many of these fruits are considered as underutilized unless the commercialization is being realized despite the fact that they have the developmental potential (Khoo et al., 2016).

Phytochemicals are the natural bioactive compounds which are found in different parts of the plant. They interplay with nutrients and dietary fibre to protect them. Phytochemical studies afford revelation and understanding of phytoconstituents. The medicinal values of the plant lie in some chemical substances that produce a definite action in the human body. Phytochemicals are the chemicals produced by the various parts of the plants. These bioactive constituents of plants are steroids, terpenoids, carotenoids, flavonoids, alkaloids, tannins, glycosides which possess anti-bacterial activity (Feroz et al., 1993). Qualitative phytochemical screening will help to understand a variety of chemical compound produced by plants. Plants have limitless ability to synthesize aromatic substances mostly phenols or their oxygen substituted derivates (Geissman, 1963).

Medicinal plants are the most exclusive source of life saving drugs for majority of the world's population. The utilization of plant cells for the production of natural or recombinant compounds of commercial interest has gained increasing attention over past decades (Canter et al., 2005). *Baccaurea courtallensis* is an evergreen tree of Phyllanthaceae family commonly known as Mootty in Malayalam and is widely distributed in the Western Ghats of India, mostly found in the moist evergreen forest of Kannada and Kerala. The edible fruits of *Baccaurea courtallensis* are used as food and also in the treatment of many infectious diseases which includes diarrhoea, dysentery, skin infection (Uduak and Kola, 2010).

Cholesterol is a sterol found in animal products such as eggs, organ meats, whole milk and its derivatives, sausages, cold cuts, skinless poultry and seafood (shrimp, oysters, shellfish, octopus, lobster), and pig meats. Epidemiological studies and clinical trials have shown that dietary cholesterol is positively associated with the risk of cardio vascular disease (CVD) through an increase in total cholesterol and LDL-C (low density lipoprotein); although the main determinant of the increase in LDL-C in humans is the consumption of saturated and trans fatty acids (Howell et al., 1997).

The American Heart Association recommends that cholesterol intake stays below 300 mg per day (Lichtenstein et al., 2006), and the National Cholesterol Education Program (NCEP) ATP III recommends the consumption of less than 200 mg per day to maximize the reduction of cholesterol by diet (Williams, 2002). In Kerala, people take diet with high amount of saturated fat, trans fat, and cholesterol like red meat, ghee and oils. This study analysis the role of phytochemicals presents in the fruit extract of *Baccaurea courtallensis* to lowering cholesterol from fatty food materials.

1.1 Objectives

The objectives of this study to evaluate the phytochemical properties of water extract of *Baccaurea courtallensis* fruit pulp and its cholesterol lowering effect on various fatty food materials.

2. Review of literature

Phytochemicals are biologically active compounds present in plants and are present naturally in plants. These are derived from various parts of plants such as leaves, flowers, seeds, roots and pulps. Plant derived compounds play a very important role in the development of various clinically useful medicines (Madhuri and Pandey, 2009).They play a vital role against number of diseases such as asthma, arthritis, cancer, etc. and has no side effects. They are considered as "man friendly medicines" since they cure diseases without causing any harm to human beings. They have formed the bases of modern drug industries (Ghurde and Malode, 2014). Now a day's these phytochemicals become more popular due to their countless uses. They are non nutritive compounds and are secondary metabolites present in small quantities in higher plants. Secondary metabolites are synthesized by the plants as part of the defence system of the plant .Nearly one third of the pharmaceuticals are of plant origin. As all the plants are able to synthesize a multitude of organic molecules or phytochemicals they are referred to as secondary metabolites (Harborne, 1982). They include alkaloids, steroids, flavonoids, terpenoids, saponins, emodins, tannins and many others (Peteros and Uy, 2010). The medicinal value of plant lies in these compounds. It is crucial to know the type of phytochemical constituents which help in knowing the biological activity that would be exhibited by that particular plant (Agbafor and Nwachukwu, 2011). Photochemical have two categories primary and secondary constituents. Primary constituents have chlorophyll, proteins, sugar and amino acid. Secondary constituents contain terpenoids and alkaloids (Wadood et al., 2013). The increasing use of plant extracts in the food, cosmetic and pharmacological industries suggest that a systematic study of plants is important. Each phytochemical have their own function. Some of them is involved in odour (terpenoids), pigmentation (tannins and quinines (Mallikaharjuna, et al., 2007). The treatment and control of diseases by the use of available plants in a locality will continue to play significant roles in medical health care implementation in developing countries of world. Nearly all cultures and civilizations from ancient times to present day have depended fully or partially on herbal medicine because of their effectiveness, affordability, low toxicity and acceptability. Due to the infectiveness of many drugs as a result of microbial resistance to available agents most especially in developing countries, more and more patients are seen in the medical centres than earlier.

Cholesterol is an integral lipid component that has been popular for its perceived negative effects on health of the humans. Public concern is more and more specifically related to meat products, especially red meat. The concern over the effects of dietary cholesterol on heart diseases and the obligatory nutritional labelling in the United States (FDA, 1993) led to the need for an efficient cholesterol determination technique. As the source of the most validated and trusted analytical methods, the association of official analytical chemists (AOAC) had adopted the first cholesterol analysis procedure for foods in 1976 (AOAC Official method 976.276) Cholesterol is waxy, fat like substance that is found in all cells of the body. Body needs some cholesterol to make hormones, vitamin D and substance that help in digesting foods. Body makes all cholesterol it needs. It is also found in food we eat. Cholesterol travels through the blood stream in small packages called lipoproteins. These packages are made of fat (lipid) on inside and protein on the outside. Two types of lipoprotein carry cholesterol throughout the body low- density lipoproteins (LDL), high density lipoproteins (HDL). Low density lipoprotein cholesterol is bad cholesterol. High LDL leads to build up of cholesterol in arteries. Arteries are blood vessels that carry blood from heart to body. HDL cholesterol is good cholesterol and carry cholesterol from all parts of the body back to the liver. Liver removes cholesterol from the body .High blood cholesterol is a condition in which we have too much cholesterol in blood. Condition has no signs and symptoms with high blood cholesterol have greater chance of getting coronary heart disease. High LDL level causes chances of increase in heart disease and high HDL level causes chances of decrease in heart disease. Cholesterol enables animal cells to dispense with a cell wall (to protect membrane integrity and cell viability) thereby allows animal cells to change shape and animal cell to move. It is the precursor for biosynthesis of steroid hormones and bile acids. Cholesterol is soluble in isopropyl myristate, ether, methanol, benzene, acetone, ethanol, chloroform, and hexane.

Baccaurea courtallensis is a genus of flowering plant belonging to the family Phyllanthaceae. It is a wild growing fruit of Western Ghats of India. It grows as an understory plant in moist evergreen forests from North Kannada to South Kerala and the adjoining parts of Tamil Nadu (western parts) up to an altitude of 900 m. It bears edible fruit but acrid in taste. The common names of *Baccaurea courtallensis* Muel. Arg. includes Mootti pazham, Mootipuli, Moootikaya, Kolikukki in Kannada, Mootipazham. The fruits are harvested by the local tribal population of the region for their medicinal value (Mohan, 2009; Dulip Daniels and Cross Bell, 2000). The tree bears tiny crimson flowers on long stalks growing on the trunk of the tree. These stalks are arranged in clusters of the trunk. Fruits hang in clusters from the trunk of the tree. The trees flower on the month of February and March, fruits are borne during the month of May and June. Little work has been reported in the chemistry of this fruit especially with regards to fruit value. Literature survey revealed lack of information on *Baccaurea courtallensis*. Its fruit is a berry consisting of an outer semi hard but fleshy rind 2-3 m thick. The cavity inside the rind is normally occupied by a single aril late seed, but, two seeds are seen occasionally. Fresh rind was found to be rich in antioxidants.

3. Hypothesis

The current research work is based on the following hypothesis

1) *Baccaurea courtallensis* extracts are rich in various phytochemical components.

2) These extract could lower cholesterol levels.

4. Materials and Methods

4.1 Study area

Kerala state covers an area of 38,863 km^2 with a population density of 859 per km^2 and spread across 14 districts. The climate is characterized by tropical wet and dry with average annual rainfall amounts to 2,817 ± 406 mm and mean annual temperature is 26.8°C (averages from 1871-2005; Krishnakumar et al., 2009). Maximum rainfall occurs from June to September mainly due to South West Monsoon and temperatures are highest in May and November.

4.2 Collection of plant material

The *Baccaurea courtallensis* fruits were collected from Erattupetta village, Thalapalam Panchayath, Meenachil Thaluk, Kottayam district, Kerala, India. It was identified taxonomically and stored.

4.3 Preparation of *Baccaurea courtallensis* fruit pulp extracts

The fruits of the plants were cleaned properly. The seeds are removed from the mature fruits and the pods were shade dried for 1-3 weeks and powdered in mechanical grinder and stored in closed vessel. Then 10 gm dried powder was mixed in 100 ml water and kept under shaker for overnight. The mixture was filtered through Whatman no 1 filter paper to precipitate and allowed to evaporate the solvents. The extract were kept in sterilized microcentrifuge tube and stored in refrigerator for further use.

4.4 Phytochemical screening

Chemical Presence or absence of certain important compounds in an extract is determined by colour reactions of the compounds with specific chemicals which act as dyes. This procedure is a simple preliminary prerequisite before going for detailed phytochemical investigation. Various tests have been conducted qualitatively to find out the presence or absence of bioactive compounds.

Chemical tests were carried out on the aqueous extract and on the powdered specimens using standard procedures to identify the constituents as described by Sofowara (1993), Trease and Evans (1989) and Harborne (1973).

4.4.1 Test for alkaloids

Two ml of plant extract was taken in a test tube and few drops of Hager's reagent were added. Yellow precipitate shows positive result for alkaloids.

4.4.2 Test for anthraquinones

Three ml of plant extract was taken in a test tube and three ml of benzene and five ml of ten percentage NH_3 were added. Formation of pink, violet or red coloration in ammonical layer detect the presence of anthraquinones.

4.4.3 Test for anthocyanins

Two ml of plant extract was taken in a test tube and two ml of 2N HCl and NH_3 were added. Formation of pinkish red to bluish violet coloration indicates the presence of anthocyanins.

4.4.4 Test for carbohydrate

Two ml of plant extract was taken in a test tube and ten ml of water, two drops of twenty percentage ethanolic α naphthol and two ml of conc. H_2SO_4 were added. Formation of reddish violet ring at the junction shows the presence of carbohydrates.

4.4.5 Test for coumarins

Two ml of extract was taken in a test tube and three ml of ten percentage NaOH was added. Formation of yellow colour gives positive result to coumarins.

4.4.6 Test for emodins

Two ml of plant extract was taken in a test tube and two ml of NH_4OH and three ml of benzene were added. Formation of red colour indicates the presence of emodins.

4.4.7 Test for flavonoids

Five ml of dilute ammonia solution were added to a portion of the plant extract followed by addition of concentrated H_2SO_4. A yellow colouration observed in each extract indicated the presence of flavonoids. The yellow colouration disappeared on standing.

4.4.8. Test for glycosides

Two ml of plant extract was taken in a test tube and two ml of chloroform and two ml of acetic acid were added. Formation of violet to blue to green coloration shows the presence of glycosides.

4.4.9 Test for leucoanthocyanins

Five ml of isoamyl alcohol taken in a test tube and five ml of plant extract was added. Turn organic layer into red detects the presence of leucoanthocyanins.

4.4.10 Test for phlobatannins

Deposition of a red precipitate when an extract of each plant sample was boiled with one percentage aqueous hydrochloric acid was taken as evidence for the presence of phlobatannins.

4.4.11 Test for proteins

One ml of plant extract was mixed with one ml of conc.H_2SO_4 in a test tube. Formation of white precipitate indicate the presence of proteins

4.4.12 Test for phenols

Few ml of the plant extract was taken in attest tube and few ml of lead acetate was added to it. Formation of white precipitate detects the presence of phenols.

4.4.13 Test for saponins

Ten ml of the extract was mixed with five ml of distilled water and shaken vigorously for a stable persistent froth. The frothing was mixed with three drops of olive oil and shaken vigorously, then observed for the formation of emulsion.

4.4.14 Test for steroids

Two ml of extract was taken in a test tube and two ml chloroform and two ml of conc.H_2SO_4 was added. Formation of reddish brown ring at the junction shows the presence of steroids.

4.4.15 Test for terpenoids

Five ml of each extract was mixed in two ml of chloroform, and concentrated H_2SO_4 (three ml) was carefully added to form a layer. A reddish brown colouration of the inter face was formed to show positive results for the presence of terpenoids.

4.5 Preparation of cholesterol samples
One gram of the sample was dissolved in one ml of chloroform and stored in brown bottle for further use (Varley, 2004).

4.6 Treatment
200 µl of extract was added to each of the sample prepared and mixed well. These are used for the periodic (24 hour interval) determination of cholesterol.

4.7 Estimation of cholesterol
The amount of cholesterol in each sample was estimated by Zak's method before and after treatment (Varley, 2004).

4.9 Statistical analysis
The survey results were analyzed and descriptive statistics were done using SPSS 12.0 (SPSS Inc., an IBM Company, Chicago, USA) and graphs were generated using Sigma Plot 7 (Systat Software Inc., Chicago, USA).

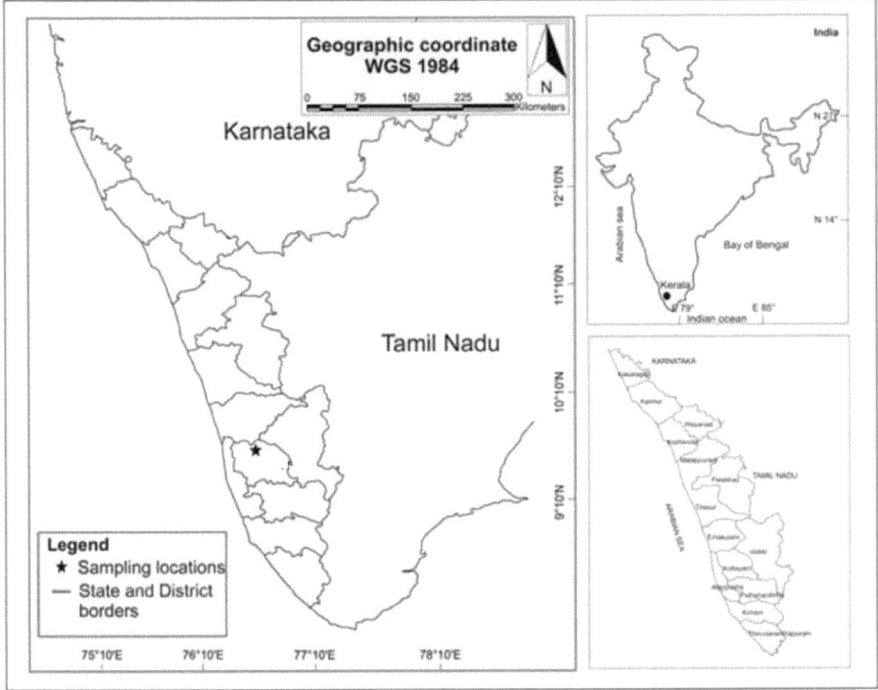

Figure 1. Map of Kerala showing the soil sample collection point. Authors own work.

Figure removed for publication – pictures retrieved from:
http://www.toxicologycentre.com/mootipazham/;
http://www.fruitipedia.com/Mootapalam_baccaurea_courtallensis.htm;
https://www.pinterest.de/pin/336503403381655120/

Figure 2. Details of the *Baccaurea courtallensis* plant with fruits.

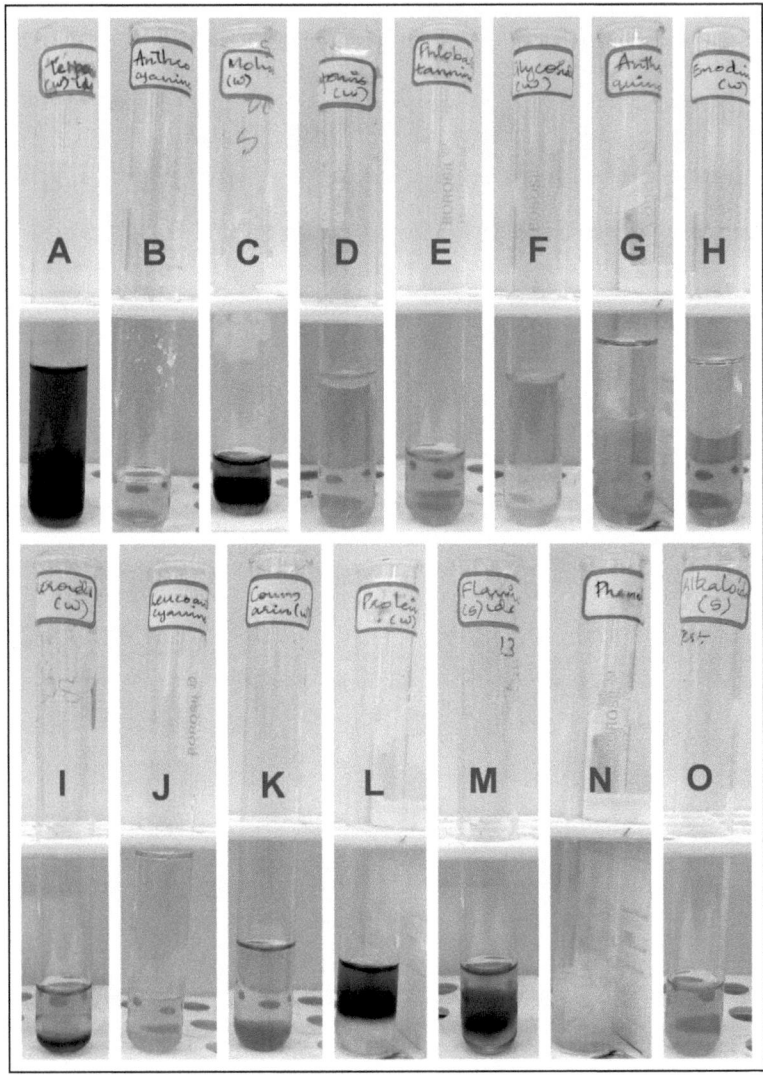

Figure 3. Details of phytochemical analysis of *Baccaurea courtallensis* fruit (water extract S1); A. terpenoids B. anthocyanins C. carbohydrates D. saponins E. phlobatanins F. glycosides G. anthraquinones H. emodins I. steroid J. leucoanthocyanin K. coumarins L. proteins M. flavonoids N. phenols O. alkaloids. Authors own image.

Table 1. Preliminary phytochemical analysis of *Baccaurea courtallensis* fruit extracts.

Sl. No	Phytoconstituents	Fruit Extract
1.	Alkaloids	+
2.	Anthrocyanin	+
3.	Anthroquinone	-
4.	Carbohydrates	+
5.	Coumarin	+
6.	Emodin	-
7.	Flavonoids	+
8.	Glycosides	-
9.	Leucoanthocyanin	+
10.	Phenols	+
11.	Phlobatannins	+
12.	Saponins	+
13.	Terpenoids	+
14.	Saponins	+
15.	Steroid	+

+ indicates presence of phytochemicals
- indicates absence of phytochemicals

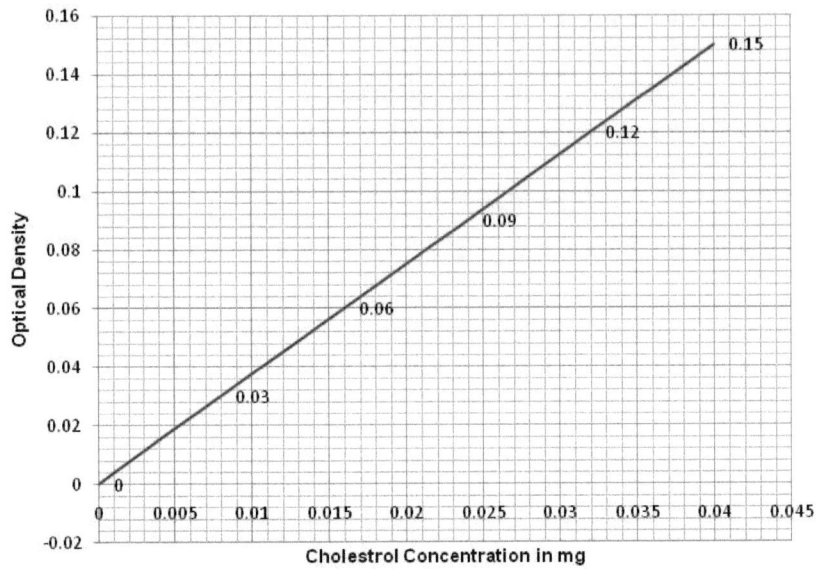

Figure 4. Standard graph for cholesterol estimation by Zak's method.

Table 2. Cholesterol estimation at different time intervals after treatment (n=3; values in mg/g sample).

Day	Substrate				
	Egg yolk	Pork Fat	Chicken Fat	Ghee	Cod liver oil
1	0.024 ± 0.01	0.022 ± 0.01	0.024 ± 0.03	0.088 ± 0.02	0.014 ± 0.03
2	0.024 ± 0.01	0.019 ± 0.01	0.024 ± 0.02	0.066 ± 0.01	0.014 ± 0.01
3	0.024 ± 0.01	0.019 ± 0.02	0.024 ± 0.01	0.024 ± 0.01	0.014 ± 0.01
4	0.022 ± 0.01	0.019 ± 0.01	0.022 ± 0.01	0.016 ± 0.01	0.014 ± 0.01
5	0.019 ± 0.01	0.016 ± 0.00	0.022 ± 0.01	0.011 ± 0.01	0.014 ± 0.00
6	0.019 ± 0.02	0.016 ± 0.01	0.022 ± 0.01	0.011 ± 0.01	0.014 ± 0.00
7	0.016 ± 0.01	0.014 ± 0.00	0.014 ± 0.01	0.011 ± 0.01	0.014 ± 0.00
8	0.016 ± 0.02	0.014 ± 0.01	0.014 ± 0.01	0.011 ± 0.01	0.014 ± 0.01
9	0.014 ± 0.00	0.014 ± 0.01	0.014 ± 0.00	0.008 ± 0.01	0.014 ± 0.01
10	0.014 ± 0.01	0.014 ± 0.01	0.014 ± 0.01	0.006 ± 0.01	0.014 ± 0.01
11	0.014 ± 0.01	0.008 ± 0.00	0.014 ± 0.00	0.006 ± 0.01	0.014 ± 0.01
12	0.011 ± 0.01	0.003 ± 0.00	0.014 ± 0.00	0.006 ± 0.00	0.014 ± 0.00

Numbers represent means ± one standard error (SE) of mean.

5. Results and discussion

In this study, preliminary phytochemical analysis of the fruit rind extracts of the *Baccaurea courtallensis* Muell-Arg showed positive tests to alkaloids, anthrocyanin, carbohydrates, coumarin, flavonoids, leucoanthocyanin, phenols, phlobatannins, saponins, terpenoids and steroids. Among extracts tested, more cholesterol lowering activity was observed on ghee followed by chicken fat. Egg yolk and pork fat shows moderate cholesterol lowering efficiency. But it was noted that there was no change in the cholesterol level of cod liver oil after treatment with the extract.

Several studies have showed that extracts of plants have a lipid lowering activity, which could prevent diseases like hyperlipidemia and cardiovascular diseases (Carvajal-Zarrabal et al., 2005; Chang et al., 2006). The extracts were able to decrease low-density lipoprotein cholesterol (LDL-c), triglycerides (TAG), total cholesterol (TC) and lipid peroxidaxion in vivo. A few of them even reported that the extract was also able to reduce very-low density lipoprotein cholesterol (VLDL-c) (Farombi and Ige, 2007; Ochani and D'Mello, 2009).

The study was conducted in vitro and found that the chemicals present in the fruit rind extract have an ability to degrade the cholesterol present in the fatty food materials. It was most effective on ghee than the other samples. There are many in vivo studies proved that photochemicals are very effective for cholesterol reduction. The cholesterol lowering effect of saponin has been demonstrated in animal and human trials (Ocakenfull and Sidhu 1990; Milgate and Roberts, 1995).

Based on the present study, we can consider the plants studied have good sources of cholesterol lowering property. The active plant extract may be further subjected to biological and pharmacological investigation for isolation of therapeutic compounds.

6. Conclusions

The study was conducted to analyses the phytochemical characteristics As a result of the study different phytochemicals were present in the species *Baccaurea courtallensis* and their hypolipidemic effect on various fatty food materials was studied. Phytochemicals are bioactive compounds that are found in different parts of the plant. Each phytochemicals have their own function. The presence of phytochemicals was tested by various tests. The different phytochemicals present in them are alkaloids, flavinoids, terpenoids, saponins, phlobatanins, coumarins, anthocyanins, leucoanthocyanins, phenols, carbohydrates. Cholesterol is an integral lipid component.

Body needs some cholesterol to make vitamin D, hormones and substances that help in digesting foods. The cholesterol sample was tested against test sample by using Zak's method on various fatty food materials such as egg yolk, ghee, chicken fat, pork fat and cod liver oil. The observations were noted colourimetrically at different time intervals. The result showed a decrease in cholesterol level within these food materials with the increase in time intervals.

Acknowledgements

The authors are grateful for the cooperation of the management of Mar Augusthinose college for necessary support. We also thank Abhilash. V. Pandiankal (Head of the department of electronics, Mar Augusthinose college, Ramapuram) for the supply of *Baccaurea courtallensis* fruit samples. Technical assistance from Binoy A Mulanthra is also acknowledged.

References

Agbafor, K. N., & Nwachukwu, N. (2011). Phytochemical analysis and antioxidant property of leaf extracts of Vitex doniana and Mucuna pruriens. *Biochemistry Research International, 2011.*

Agbafor, K. N., & Nwachukwu, N. (2011). Phytochemical analysis and antioxidant property of leaf extracts of Vitex doniana and Mucuna pruriens. *Biochemistry Research International, 2011.*

Canter, P. H., Thomas, H., & Ernst, E. (2005). Bringing medicinal plants into cultivation: opportunities and challenges for biotechnology. *TRENDS in Biotechnology, 23*(4), 180-185.

Carvajal-Zarrabal, O., Waliszewski, S. M., Barradas-Dermitz, D. M., Orta-Flores, Z., Hayward-Jones, P. M., Nolasco-Hipólito, C., & Trujillo, P. R. (2005). The consumption of Hibiscus sabdariffa dried calyx ethanolic extract reduced lipid profile in rats. *Plant Foods for Human Nutrition (Formerly Qualitas Plantarum), 60*(4), 153-159.

Chang, Y. C., Huang, K. X., Huang, A. C., Ho, Y. C., & Wang, C. J. (2006). Hibiscus anthocyanins-rich extract inhibited LDL oxidation and oxLDL-mediated macrophages apoptosis. *Food and Chemical Toxicology, 44*(7), 1015-1023.

Dulip Daniels, A. E., & Cross Bell, D. (2000). Star of the forest. *The Hindu, Chennai, India.*

Farombi, E. O., & Ige, O. O. (2007). Hypolipidemic and antioxidant effects of ethanolic extract from dried calyx of Hibiscus sabdariffa in alloxan-induced diabetic rats. *Fundamental & clinical pharmacology, 21*(6), 601-609.

Farombi, E. O., & Ige, O. O. (2007). Hypolipidemic and antioxidant effects of ethanolic extract from dried calyx of Hibiscus sabdariffa in alloxan-induced diabetic rats. *Fundamental & clinical pharmacology, 21*(6), 601-609.

Feroz, M., Ahmad, R., Sindhu, S. T. A. K., & Shahbaz, A. M. (1993). Antifungal activities of saponins from indigenous plant roots. *Pakistan Veterinary Journal 13*(1), 4-4.

Geissman, T. A. (1963). Flavonoid compounds, tannins, lignins and related compounds. *Pyrrole pigments, isoprenoid compounds and phenolic plant constituents, 9*, 265.

Geisssman, T.A., 1963. Flavonoid compounds, Tannins, Lignins and related compounds, In M. Florkin and Stotz (Ed), Pyrrole Pigments, Isoprenoid compounds and phenolic plant constituents Vol. 9. 265.

Ghurde, O. M., & Malode, S. N. (2014). Phytochemical screening and assessment of biomolecules compounds in Scilla hyacinthina (Roth) Macbr. Bulb. *J. GlobalBiosci*, *3*(5), 866-871.

Harborne, J. B. (1973). Introduction to Ecological Bio Chemistry, Second ED, Academic Press, New York, NY.

Harborne, J. B. (1973). Phenolic compounds. In *Phytochemical methods* (pp. 33-88). Springer Netherlands.

Howell, W. H., McNamara, D. J., Tosca, M. A., Smith, B. T., & Gaines, J. A. (1997). Plasma lipid and lipoprotein responses to dietary fat and cholesterol: a meta-analysis. *The American Journal of Clinical Nutrition*, *65*(6), 1747-1764.

Howell, W. H., McNamara, D. J., Tosca, M. A., Smith, B. T., & Gaines, J. A. (1997). Plasma lipid and lipoprotein responses to dietary fat and cholesterol: a meta-analysis. *The American journal of clinical nutrition*, *65*(6), 1747-1764.

Khoo, H. E., Azlan, A., Kong, K. W., & Ismail, A. (2016). Phytochemicals and medicinal properties of indigenous tropical fruits with potential for commercial development. *Evidence-Based Complementary and Alternative Medicine*, *2016*.

Krishnakumar, K. N., Rao, G. P., & Gopakumar, C. S. (2009). Rainfall trends in twentieth century over Kerala, India. *Atmospheric environment*, *43*(11), 1940-1944.

Lichtenstein, A. H., Appel, L. J., Brands, M., Carnethon, M., Daniels, S., Franch, H. A., & Karanja, N. (2006). Diet and lifestyle recommendations revision 2006. *Circulation*, *114*(1), 82-96.

Madhuri, S., & Pandey, G. (2009). Some anticancer medicinal plants of foreign origin. *Current Science*, *96*(6), 779-783.

Mallikarjuna, N., Sharma, H. C., & Upadhyaya, H. D. (2007). Exploitation of wild relatives of pigeonpea and chickpea for resistance to Helicoverpa armigera. *Jounal of SAT Agricultural Research*, *3*(1), 4.

Mallikharjuna, P. B., Rajanna, L. N., Seetharam, Y. N., & Sharanabasappa, G. K. (2007). Phytochemical studies of Strychnos potatorum Lf-A medicinal plant. *Journal of Chemistry*, *4*(4), 510-518.

Mohan, S. (2009). Fatty acid composition of *Baccaurea courtallensis* Muell. Arg seed oil: An endemic species of Western Ghats, India. *Journal of the American Oil Chemists' Society*, *86*(10), 1017-1019.

Mohan, S. (2009). Fatty acid composition of Baccaurea courtallensis Muell. Arg seed oil: An endemic species of western Ghats, India. *Journal of the American Oil Chemists' Society*, *86*(10), 1017-1019.

Oakenfull, D., & Sidhu, G. S. (1990). Could saponins be a useful treatment for hypercholesterolaemia?. *European Journal of Clinical Nutrition*, *44*(1), 79-88.

Ochani, P. C., & D'Mello, P. (2009). Antioxidant and antihyperlipidemic activity of Hibiscus sabdariffa Linn. leaves and calyces extracts in rats. *Indian Journal of Experimental Biology*, 47(4), 276–282.

Ochani, P. C., & D'Mello, P. (2009). Antioxidant and antihyperlipidemic activity of *Hibiscus sabdariffa* Linn. leaves and calyces extracts in rats.

Peteros, N. P., & Uy, M. M. (2010). Antioxidant and cytotoxic activities and phytochemical screening of four Philippine medicinal plants. *Journal of Medicinal Plants Research*, *4*(5), 407-414.

Sofowora, A. (1993). Recent trends in research into African medicinal plants. *Journal of Ethnopharmacology*, *38*(2-3), 197-208.

Trease, G. E., & Evans, W. C. Pharmacognosy. 1989. *Bailliere Tindall, London*, 45-50.

Uduak, A. E., & Kola, K. A. (2010). Antimicrobial activities of some Euphorbiaceae plants used in the traditional medicine of Akwa Ibom State of Nigeria. *Ethnobotanical Leaflets*, *2010*(6), 2.

Varley, H. (2004). Practical clinical Biochemistry, 4th edition, 2004, Pg 313-17

Varley, H. (2004). Practical clinical Biochemistry, 4th edition, Heinemann Medical, UK.

Wadood, A., Ghufran, M., Jamal, S. B., Naeem, M., Khan, A., Ghaffar, R., & Asnad, C. (2013). Phytochemical analysis of medicinal plants occurring in local area of Mardan. *Biochem Anal Biochem*, *2*(4), 1-4.

Wadood, A., Ghufran, M., Jamal, S. B., Naeem, M., Khan, A., Ghaffar, R., & Asnad, C. (2013). Phytochemical analysis of medicinal plants occurring in local area of Mardan. *Biochem Anal Biochem*, *2*(4), 1-4.

Williams, L. (2002). Third report of the National Cholesterol Education Program (NCEP) expert panel on detection, evaluation, and treatment of high blood cholesterol in adults (Adult Treatment Panel III) final report. *Circulation, 106*(25), 3143-3143.

YOUR KNOWLEDGE HAS VALUE

- We will publish your bachelor's and
  master's thesis, essays and papers

- Your own eBook and book -
  sold worldwide in all relevant shops

- Earn money with each sale

Upload your text at www.GRIN.com
and publish for free